# The Mystery of the Charity
## of Charles Péguy

By the same author

\*

*Geoffrey Hill*

# The Mystery of the Charity
# of Charles Péguy

New York
OXFORD UNIVERSITY PRESS
1984

Copyright © 1983 by Geoffrey Hill
First published in Great Britain by Agenda Editions
and André Deutsch Limited, 1983
First published in the United States
by Oxford University Press, New York, 1984
First issued as an Oxford University Press paperback, 1984

Library of Congress Cataloging in Publication Data
Hill, Geoffrey.
The mystery of the charity of Charles Péguy.
1. Péguy, Charles, 1873–1914, in fiction,
drama, poetry, etc.   I. Title.
PR6015.I4735M9   1984   821'.914   84-14782
ISBN 0-19-503514-3
ISBN 0-19-503515-1 (pbk.)

Printing (last digit): 9 8 7 6 5 4 3 2 1

Printed in the United States of America

# Acknowledgements

Prior to its appearance in print *The Mystery of the Charity of Charles Péguy* was broadcast on BBC Radio 3 on February 1, 1983. It was read by Paul Webster. The producer was Fraser Steel.

The poem appeared in the *Times Literary Supplement*, February 4, 1983, and in *The Paris Review*, Summer, 1983.

In loving memory

of

SARAH ANN HANDS

1869–1967

The Mystery of the Charity
of Charles Péguy

*Nous sommes les derniers. Presque les après-derniers. Aussitôt après nous commence un autre âge, un tout autre monde, le monde de ceux qui ne croient plus à rien, qui s'en font gloire et orgueil.*

Charles Péguy

# 1

Crack of a starting-pistol. Jean Jaurès
dies in a wine-puddle. Who or what stares
through the café-window crêped in powder-smoke?
The bill for the new farce reads *Sleepers Awake.*


History commands the stage wielding a toy gun,
rehearsing another scene. It has raged so before,
countless times; and will do, countless times more,
in the guise of supreme clown, dire tragedian.


In Brutus' name martyr and mountebank
ghost Caesar's ghost, his wounds of air and ink
painlessly spouting. Jaurès' blood lies stiff
on menu-card, shirt-front and handkerchief.


Did Péguy kill Jaurès? Did he incite
the assassin? Must men stand by what they write
as by their camp-beds or their weaponry
or shell-shocked comrades while they sag and cry?


Would Péguy answer—stubbornly on guard
among the *Cahiers,* with his army cape
and steely pince-nez and his hermit's beard,
brooding on conscience and embattled hope?

Truth's pedagogue, braving an entrenched class
of fools and scoundrels, children of the world,
his eyes caged and hostile behind glass—
still Péguy said that Hope is a little child.

Violent contrariety of men and days; calm
juddery bombardment of a silent film
showing such things: its canvas slashed with rain
and St. Elmo's fire. Victory of the machine!

The brisk celluloid clatters through the gate;
the cortège of the century dances in the street;
and over and over the jolly cartoon
armies of France go reeling towards Verdun.


2


Rage and regret are tireless to explain
stratagems of the out-manoeuvred man,
the charge and counter-charge. You know the drill,
raw veteran, poet with the head of a bull.

Footslogger of genius, skirmisher with grace
and ill-luck, sentinel of the sacrifice,
without vantage of vanity, though mortal-proud,
defend your first position to the last word.

The sun-tanned earth is your centurion;
you are its tribune. On the hard-won
high places the old soldiers of old France
crowd like good children wrapped in obedience

and sleep, and ready to be taken home.
Whatever that vision, it is not a child's;
it is what a child's vision can become.
Memory, Imagination, harvesters of those fields,

our gifts are spoils, our virtues epitaphs,
our substance is the grass upon the graves.
"Du calme, mon vieux, du calme." How studiously
one cultivates the sugars of decay,

pâtisserie-tinklings of angels "'sieur-'dame,"
the smile of the dead novice in its plush frame,
while greed and disaffection are ingrained
like chalk-dust in the ranklings of the mind.

"Rather the Marne than the *Cahiers.*" True enough,
you took yourself off. Dying, your whole life
fell into place. "'Sieurs-'dames, this is the wall
where he leaned and rested, this is the well

15

from which he drank." Péguy, you mock us now.
History takes the measure of your brow
in blank-eyed bronze, brave mediocre work
of *Niclausse, sculpteur,* cornered in the park

among the stout dogs and lame patriots
and all those ghosts, far-gazing in mid-stride,
rising from where they fell, still on parade,
covered in glory and the blood of beetroots.


3


Vistas of richness and reward. The cedar
uprears its lawns of black cirrus. You have found
hundred-fold return though in the land
of exile. You are Joseph the Provider;

and in the fable this is your proper home;
three sides of a courtyard where the bees thrum
in the crimped hedges and the pigeons flirt
and paddle, and sunlight pierces the heart-

shaped shutter-patterns in the afternoon,
shadows of fleurs-de-lys on the stone floors.
Here life is labour and pastime and orison
like something from a simple book of hours;

and immortality, your measured task,
inscribes its antique scars on the new desk
among your relics, bits of ivory quartz
and dented snuffbox won at Austerlitz.


The proofs pile up; the dead are made alive
to their posthumous fame. Here is the archive
of your stewardship; here is your true domaine,
its fields of discourse ripening to the Marne.


Chateau de Trie is yours, Chartres is yours,
and the carved knight of Gisors with the hound;
Colombey-les-deux-Eglises; St. Cyr's
cadres and echelons are yours to command.


Yours is their dream of France, militant-pastoral:
musky red gillyvors, the wicker bark
of clematis braided across old brick
and the slow chain that cranks into the well


morning and evening. It is Domrémy
restored; the mystic strategy of Foch
and Bergson with its time-scent, dour panache
deserving of martyrdom. It is an army

of poets, converts, vine-dressers, men skilled
in wood or metal, peasants from the Beauce,
terse teachers of Latin and those unschooled
in all but the hard rudiments of grace.

Such dreams portend, the dreamer prophesies,
is this not true? Truly, if you are wise,
deny such wisdom; bid the grim bonne-femme
defend your door: "M'sieur is not at home."


4


This world is different, belongs to them—
the lords of limit and of contumely.
It matters little whether you go tamely
or with rage and defiance to your doom.

This is your enemies' country which they took
in the small hours an age before you woke,
went to the window, saw the mist-hewn
statues of the lean kine emerge at dawn.

Outflanked again, too bad! You still have pride,
haggard obliquities: those that take remorse
and the contempt of others for a muse,
bound to the alexandrine as to the *Code*

*Napoléon.* Thus the bereaved soul returns
upon itself, grows resolute at chess,
in war-games hurling dice of immense loss
into the breach; thus punitively mourns.


This is no old Beauce manoir that you keep
but the rue de la Sorbonne, the cramped shop,
its unsold *Cahiers* built like barricades,
its fierce disciples, disciplines and feuds,


the camelot-cry of "sticks!" As Tharaud says,
"all through your life the sound of broken glass."
So much for Jaurès murdered in cold pique
by some vexed shadow of the belle époque,


some guignol strutting at the window-frame.
But what of you, Péguy, who came to "exult,"
to be called "wolfish" by your friends? The guilt
belongs to time; and you must leave on time.


Jaurès was killed blindly, yet with reason:
"let us have drums to beat down his great voice."
So you spoke to the blood. So, you have risen
above all that and fallen flat on your face

## 5

among the beetroots, where we are constrained
to leave you sleeping and to step aside
from the fleshed bayonets, the fusillade
of red-rimmed smoke like stubble being burned;

to turn away and contemplate the working
of the radical soul—instinct, intelligence,
memory, call it what you will—waking
into the foreboding of its inheritance,

its landscape and inner domain; images
of earth and grace. Across Artois the rois-mages
march on Bethlehem; sun-showers fall
slantwise over the kalefield, the canal.

Hedgers and ditchers, quarrymen, thick-shod
curés de campagne, each with his load,
shake off those cares and burdens; they become,
in a bleak visionary instant, seraphim

looking towards Chartres, the spired sheaves,
stone-thronged annunciations, winged ogives
uplifted and uplifting from the winter-gleaned
furrows of that criss-cross-trodden ground.

Or say it is Pentecost: the hawthorn-tree,
set with coagulate magnified flowers of may,
blooms in a haze of light; old chalk-pits brim
with seminal verdure from the roots of time.


Landscape is like revelation; it is both
singular crystal and the remotest things.
Cloud-shadows of seasons revisit the earth,
odourless myrrh borne by the wandering kings.


Happy are they who, under the gaze of God,
die for the "terre charnelle," marry her blood
to theirs, and, in strange Christian hope, go down
into the darkness of resurrection,


into sap, ragwort, melancholy thistle,
almondy meadowsweet, the freshet-brook
rising and running through small wilds of oak,
past the elder-tump that is the child's castle.


Inevitable high summer, richly scarred
with furze and grief; winds drumming the fame
of the tin legions lost in haystack and stream!
Here the lost are blest, the scarred most sacred:

odd village workshops grimed and peppercorned
in a dust of dead spiders, paper-crowned
sunflowers with the bleached heads of rag dolls,
brushes in aspic, clay pots, twisted nails;


the clinking anvil and clear sheepbell-sound,
at noon and evening, of the angelus;
coifed girls like geese, labourers cap in hand,
and walled gardens espaliered with angels;


solitary bookish ecstasies, proud tears,
proud tears, for the forlorn hope, the guerdon
of Sedan, "oh les braves gens!", English Gordon
stepping down sedately into the spears.


Patience hardens to a pittance, courage
unflinchingly declines into sour rage,
the cobweb-banners, the shrill bugle-bands
and the bronze warriors resting on their wounds.


These fatal decencies, they make us lords
over ourselves: familial debts and dreads,
keepers of old scores, the kindly ones
telling their beady sous, the child-eyed crones

who guard the votive candles and the faint
invalid's night-light of the sacrament,
a host of lilies and the table laid
for early mass from which you stood aside

to find salvation, your novena cleaving
brusquely against the grain of its own myth,
its truth and justice, to a kind of truth,
a justice hard to justify. "Having

spoken his mind he'd a mind to be silent."
But who would credit that, that one talent
dug from the claggy Beauce and returned to it
with love, honour, suchlike bitter fruit?

## 6

To dispense, with justice; or, to dispense
with justice. Thus the catholic god of France,
with honours all even, honours all, even
the damned in the brazen Invalides of Heaven.

Here there should be a section without words
for military band alone: "Sambre et Meuse,"
the "Sidi Brahim" or "Le Roi s'Amuse";
white gloves and monocles and polished swords

and Dreyfus with his buttons off, chalk-faced
but standing to attention, the school prig
caught in some act and properly disgraced.
A puffy satrap prances on one leg

to snap the traitor's sword, his ordered rage
bursting with "cran et gloire" and gouts of rouge.
The chargers click and shiver. There is no stir
in the drawn ranks, among the hosts of the air,

all draped and gathered by the weird storm-light
cheap wood-engravings cast on those who fought
at Mars-la-Tour, Sedan; or on the men
in the world-famous stories of Jules Verne

or nailed at Golgotha. Drumrap and fife
hit the right note: "A mort le Juif! Le Juif
à la lanterne!" Serenely the mob howls,
its silent mouthings hammered into scrolls

torn from *Apocalypse.* No wonder why
we fall to violence out of apathy,
redeemed by falling and restored to grace
beyond the dreams of mystic avarice.

But who are "we," since history is law,
clad in our skins of silver, steel and hide,
or in our rags, with rotten teeth askew,
heroes or knaves as Clio shall decide?


"We" are crucified Pilate, Caiaphas
in his thin soutane and Judas with the face
of a man who has drunk wormwood. We come
back empty-handed from Jerusalem


counting our blessings, honestly admire
the wrath of the peacemakers, for example
Christ driving the money-changers from the temple,
applaud the Roman steadiness under fire.


We are the occasional just men who sit
in gaunt self-judgment on their self-defeat,
the élite hermits, secret orators
of an old faith devoted to new wars.


We are "embusqués," having no wounds to show
save from the thorns, ecstatic at such pain.
Once more the truth advances; and again
the metaphors of blood begin to flow.

7

Salute us all, Christus with your iron
garlands of poppies and ripe carrion.
No, sleep where you stand; let some boy-officer
take up your vigil with your dungfork spear.


What vigil is this, then, among the polled
willows, cart-shafts uptilted against skies,
translucent rain at jutting calvaries;
on paths that are rutted and broken-walled?


What is this relic fumbled with such care
by mittened fingers in dugout or bomb-
tattered, jangling estaminet's upper room?
The incense from a treasured tabatière,


you watchmen at the Passion. Péguy said
"why do I write of war? Simply because
I have not been there. In time I shall cease
to invoke it." We still dutifully read


"heureux ceux qui sont morts." Drawn on the past
these presences endure; they have not ceased
to act, suffer, crouching into the hail
like labourers of their own memorial

or those who worship at its marble rote,
their many names one name, the common "dur"
built into duration, the endurance of war;
blind Vigil herself, helpless and obdurate.

And yet what sights: Saul groping in the dust
for his broken glasses, or the men far-gone
on the road to Emmaus who saw the ghost.
Commit all this to memory. The line

falters, reforms, vanishes into the smoke
of its own unknowing; mother, dad,
gone in that shell-burst, with the other dead,
"pour la patric," according to the book.

8

Dear lords of life, stump-toothed, with ragged breath,
throng after throng cast out upon the earth,
flesh into dust, who slowly come to use
dreams of oblivion in lieu of paradise,

push on, push on!—through struggle, exhaustion,
indignities of all kinds, the impious Christian
oratory, "vos morituri," through berserk fear,
laughing, howling, "servitude et grandeur"

in other words, in nameless gobbets thrown
up by the blast, names issuing from mouths
of the dying, with their dying breaths.
But rest assured, bristly-brave gentlemen

of Normandie and Loire. Death does you proud,
every heroic commonplace, "Amor,"
"Fidelitas," polished like old armour,
stamped forever into the featureless mud.

Poilus and sous-officiers who plod
to your lives' end, name your own recompense,
expecting nothing but the grace of France,
drawn to her arms, her august plenitude.

The blaze of death goes out, the mind leaps
for its salvation, is at once extinct;
its last thoughts tetter the furrows, distinct
in dawn twilight, caught on the barbed loops.

Whatever strikes and maims us it is not
fate, to our knowledge. En avant, Péguy!
The irony of advancement. Say "we
possess nothing; try to hold on to that."

## 9

There is an ancient landscape of green branches—
true tempérament de droite, you have your wish—
crosshatching twigs and light, goldfinches
among the peppery lilac, the small fish


pencilled into the stream. Ah, such a land
the Ile de France once was. Virelai and horn
wind through the meadows, the dawn-masses sound
fresh triumphs for our Saviour crowned with scorn.


Good governors and captains, by your leave,
you also were sore-wounded but those wars
are ended. Iron men who bell the hours,
marshals of porte-cochère and carriage-drive,


this is indeed perfection, this is the heart
of the mystère. Yet one would not suppose
Péguy's "defeat," "affliction," your lost cause.
Old Bourbons view-hallooing for regret


among the cobwebs and the ghostly wine,
you dream of warrior-poets and the Meuse
flowing so sweetly; the androgynous Muse
your priest-confessor, sister-châtelaine.

How the mood swells to greet the gathering storm!
The chestnut trees begin to thresh and cast
huge canisters of blossom at each gust.
Coup de tonnerre! Bismarck is in the room!

Bad memories, seigneurs? Such wraiths appear
on summer evenings when the gnat-swarm spins
a dying moment on the tremulous air.
The curtains billow and the rain begins

its night-long vigil. Sombre heartwoods gleam,
the clocks replenish the small hours' advance
and not a soul has faltered from its trance.
"Je est un autre," that fatal telegram,

floats past you in the darkness, unreceived.
Connoisseurs of obligation, history
stands, a blank instant, awaiting your reply:
"If we but move a finger France is saved!"

10

Down in the river-garden a grey-gold
dawnlight begins to silhouette the ash.
A rooster wails remotely over the marsh
like Mr. Punch mimicking a lost child.

At Villeroy the copybook lines of men
rise up and are erased. Péguy's cropped skull
dribbles its ichor, its poor thimbleful,
a simple lesion of the complex brain.


Woefully battered but not too bloody,
smeared by fraternal root-crops and at one
with the fritillary and the veined stone,
having composed his great work, his small body,


for the last rites of truth, whatever they are,
or the Last Judgment which is much the same,
or Mercy, even, with her tears and fire,
he commends us to nothing, leaves a name


for the burial-detail to gather up
with rank and number, personal effects,
the next-of-kin and a few other facts;
his arm over his face as though in sleep


or to ward off the sun: the body's prayer,
the tribute of his true passion, for Chartres
steadfastly cleaving to the Beauce, for her,
the Virgin of innumerable charities.

"Encore plus douloureux et doux." Note how
sweetness devours sorrow, renders it again,
turns to affliction each more carnal pain.
Whatever is fulfilled is now the law


where law is grace, that grace won by inches,
inched years. The men of sorrows do their stint,
whose golgothas are the moon's trenches,
the sun's blear flare over the salient.


J'accuse! j'accuse!—making the silver prance
and curvet, and the dust-motes jig to war
across the shaky vistas of old France,
the gilt-edged maps of Strasbourg and the Saar.


Low tragedy, high farce, fight for command,
march, counter-march, and come to the salute
at every hole-and-corner burial-rite
bellowed with hoarse dignity into the wind.


Take that for your example! But still mourn,
being so moved: éloge and elegy
so moving on the scene as if to cry
"in memory of those things these words were born."

2.1.   poet with the head of a bull./*Poetry*, a tapestry by Jean Lurçat, depicts the twelve signs of the zodiac and a poet with the head of a bull.

2.7.   Rather the Marne than the *Cahiers.*/adapts a phrase from a review-article by P. McCarthy, TLS, 16th June, 1978, p. 675.

4.1.   the lords of limit/The phrase is Auden's, from an early poem "Now from my window-sill I watch the night." See *The English Auden*, edited by Edward Mendelson (London, 1977), pp. 115–16.

4.6.   the camelot-cry of "sticks!"/*Les camelots du roi* was a right-wing, anti-Dreyfusard organization, prominent in the street-battles of the period.

4.6.   As Tharaud says,/Daniel Halévy, *Péguy and "Les Cahiers de la Quinzaine,"* translated from the French by Ruth Bethell (London, 1946), p. 171: "Always, all through his life, this sound of broken glass, to use Tharaud's expression."

5.8.   die for the "terre charnelle,"/Charles Péguy, *Ève* (1913): "—Heureux ceux qui sont morts pour la terre charnelle,/Mais pourvu que se fût dans une juste guerre."

9.1.   true tempérament de droite, you have your wish—/See Robert Speaight, *Georges Bernanos* (London, 1973), for "what Jacques Maritain has called a *tempérament de droite.*" See also pp. 17–18 for Speaight's view of the great similarities, as well as the great differences, between Bernanos and Péguy.

9.8.   "Je est un autre," that fatal telegram,/Arthur Rimbaud, *Lettre à Georges Izambard*, May 1871: "vous ne comprendrez pas du tout, et je ne saurais presque vous expliquer . . . *Je* est un autre . . . "

10.7.   Encore plus douloureux et doux./from a quatrain by Charles Péguy.

10.11.   in memory of those things these words were born./adapts a sentence from Marcel Raymond, *From Baudelaire to Surrealism* (London, 1961), p. 190: referring to Péguy's "Présentation de la Beauce à Notre-Dame de Chartres."

# CHARLES PÉGUY

Charles Péguy was born in 1873 into a family of barely literate peasants, to whom he subsequently devoted much eloquent homage, and died, an ageing infantry lieutenant of the Reserve, on the first day of the first Battle of the Marne in September 1914. He was a son of the people, of "l'ancienne France," one of the last of that race as he conceived of it. His reputation, such as it was during his lifetime, was confined to a small intellectual élite: the few hundred readers of *Les Cahiers de la Quinzaine,* which he founded in 1900, and the dozen or so who attended the Thursday meetings in his little bookshop, the "Boutique des Cahiers," in the shadow of the Sorbonne. A man of the most exact and exacting probity, accurate practicality, in personal and business relations, a meticulous reader of proof, he was at the same time moved by violent emotions and violently afflicted by mischance. Like others similarly wounded, he was perhaps smitten by the desirability of suffering. "Fils de vaincu, il est attiré par les défaites": such is the suggestion of Simone Fraisse; and a further remark, quoted by Halévy, "Always, all through his life, this sound of broken glass," felicitously evokes a variety of painful scenes: from the street-battles, the riotous fringe of "L'Affaire Dreyfus" (that extraordinary collision of two kinds of patriotism, the one cynical, reactionary, the other regenerative and sacrificial) to the harsh severing of old alliances and friendships in the years that followed. A staunchly committed Dreyfusard, Péguy was an admirer of the great socialist deputy Jean Jaurès throughout the period of the "Affair." By 1914 he was calling for his blood: figuratively, it must be said; though a young madman, who may or may not have been over-susceptible to metaphor, almost immediately shot Jaurès through the head.

Péguy had become a socialist during his college days and remained one, though of an increasingly eccentric cast of thought and speech. T. Stearns Eliot, M.A. (Harvard), who made reference to Péguy's life and work in a series of university extension lectures in 1916, noted that he "illustrates nationalism and neo-Catholicism as well as socialism," and treated his ideas in close association with those of Georges Sorel. It has been said that "Péguy's socialism re-emerged as the national-socialism of Barrès and Sorel"; but fascism, in whatever form, is a travesty of Péguy's true faith and position. He did not, in the end, have a great deal in common with Sorel; quarrelled with him; was certainly not anti-semitic.

His brave and timely death in a beetroot field by the Marne transformed this much-snubbed irascible man into the kind of figure-in-profile for whom church and civic dignitaries turn out in force, whose "essential idea" even Ministers of Education may safely extol.

No one knows for certain whether he did, or did not, receive the sacrament on the Feast of the Assumption, shortly before he was killed. Estranged from the Church for a number of years, first by his militant socialist principles, then by the consequences of a secular marriage, he had, in 1908, rediscovered the solitary ardours of faith but not the consolations of religious practice. He remained self-excommunicate but adoring; his devotion most doggedly expressed in those two pilgrimages undertaken on foot, in June 1912 and July 1913, from Paris to the Cathedral of Notre Dame de Chartres. The purpose of his first journey, as a tablet in the Cathedral duly records, was to entrust his children to Our Lady's care.

There is still a "Boutique des Cahiers," a handy stone's throw from the Sorbonne. Its appearance, at least on the outside, seems remarkably unchanged from that preserved by photographs taken in 1902 and 1913 except, of course, that there is now a plaque on the wall. John Middleton Murry, in his autobiography, affords "a glimpse . . . through the windows of his little shop . . . of a man with a pince-nez set awry on his nose, tying up a parcel: that was Charles Péguy. I admired him, and admire him still." In this vignette we too glimpse something of the tragi-comic battered élan of Péguy's life. Murry's final cadence is without reservation, and I like him for such an expression of outright admiration. Péguy, stubborn rancours and mishaps and all, is one of the great souls, one of the great prophetic intelligences, of our century. I offer *The Mystery of the Charity of Charles Péguy* as my homage to the triumph of his "defeat."